Invent, Innovate, Create To Become Financially Free

Turn Your Ideas Into Money Machine

George K.

Table of Contents

Introduction

I want to thank you and congratulate you for purchasing the book, *"Invent, Innovate, Create to become Financially Free: Turn your Ideas into Money Machine"*.

This book contains proven steps and strategies on how to bring out the entrepreneur in you. We will define what it means to be creative and inventive while having an eye for business. The book will guide you through the many ways you can let your creative juices flow and translate your ideas into practical products that can be sold to the market.

If you've always had the gift for innovation, the sort of person who enjoys time spent tinkering with things and finding ways to make them better, then this book is for you. In this book, you will know how you can nurture that gift and bring out the inner inventor in you. And most important of all, monetize that talent by encouraging you to produce your own innovative products.

The value of creating prototypes, basic three dimensional renditions of product design sketches, is also discussed in detail. Prototypes allow inventors to work with their idea as a tangible product. They can test, check and refine a prototype into the best version of the product before it is sold to the market. Every serious inventor will tell you prototypes are absolutely necessary in the process of invention.

If you are looking to be encouraged as an inventor and entrepreneur then this book will give you that little push to get you on your way to translating your ideas into inventive, innovative, and creative products to sell to the market.

Thanks again for purchasing this book. I hope you enjoy it!

Chapter 1: Ideas to Invention

All inventions stem from an idea. Every product that we use on a daily basis was incubated first as an idea in the mind of its inventor. They are the result of light-bulb moments that were transformed into tangible functional products through the process of invention.

The iPhone was first an idea before Steve Jobs decided to go through with the process of inventing it. Facebook was first an idea before Mark Zuckerberg decided to build it and grow it into a billion-member strong social media website. Most of the gadgets and products we use on a daily basis were first ideas before their originators took it upon themselves to transform them into practical tools and gadgets we use today.

One of the most creative minds and prolific inventors of his time was Leonardo Da Vinci. His mind had seemingly no limits for creative invention. His capacity to imagine all sorts of gadgets and contraptions was limitless. His inventiveness was way ahead of its time. He was the first person to think of helicopters centuries before they were invented. He first thought of the possibility of armored cars, the crossbow, the parachute, the clock, scuba gear and so many other products and gadgets that were all way ahead of his time. He was the quintessential a 'Renaissance Man'.

Although not everyone can be as creatively prolific as Leonardo Da Vinci, there are a lot of people with the same gift for invention and innovation. The kind of person capable of coming up with all sorts of creative ideas that can be turned into useful inventions. If you believe you have the same gift for invention as Leonardo Da Vinci, I encourage you to nurture that gift. Now is the time to be an inventor and entrepreneur. The 21st century rewards the creative thinker and doer capable of translating ideas to useful products that benefit people.

And given the right product, massive success awaits the inventor and entrepreneur able to go through the process of inventing new products. Just look at Steve Jobs and the range of Apple products he and his team created. Because Steve Jobs willed himself to transform idea into reality, Apple is now worth billions of dollars. Steve Jobs' inventive and entrepreneurial spirit was rewarded by a global market that developed an intense and loyal appreciation for his Apple products.

Like Da Vinci and all the other inventors that preceded him, make sure you get in the habit of writing down your ideas. They are the kernels of gold that will lead you to your success. Because ideas are so fleeting, recording them allows you to refer to them again in the future and build on initial concepts. As soon as a new idea pops in your mind, write it down, record it so that it is saved for posterity. Then when you have the time, revisit the idea you have written down and try to expand it. Develop it into something more concrete and practical. If necessary, draw the product that is forming in your mind. Create an initial sketch to give you a visual of the product. Keep on making improvements until you are fully satisfied with the product that you have conceptualized.

There will come a time where you will want to transition a fully ripened idea into the invention phase. Invention is the stage where inventors transform an idea into reality. All inventors, entrepreneurs and designers, the likes of Steve Jobs, Elon Musk, and Bill Gates, to name a few, are motivated to build things. They think to create. They create to sell.

When you have gone through the process of developing the concept model of a product or gadget and you have the design sketches to prove it. Then go ahead and transition it to invention phase. All ideas deserve to see the light of reality. Especially true for those ideas you have spent valuable time, energy and resource building in your mind.

INVENTION: THE PRACTICAL WAY

Like we mentioned earlier, almost all inventors see the value in writing down their ideas on paper. Most inventions were first notes, drawings or scribbles on notepads before they were built into functional products. To invent is to make a commitment to create. It means finding time to build your product. It means allotting a certain amount of financial resource for you to be able to gather all relevant materials that will allow you to create it from scratch. To invent is to give birth to an idea.

All inventors go through a trial and error process to check the functionality and integrity of an invention. Like I mentioned earlier, to be able to test an invention, prototypes or basic models are often built. They are made up of very basic every day materials and are meant to be crude renditions of a design.

Inventors either work alone, in pairs or in groups. When working with other people they go through the process of brainstorming. When people brainstorm they create the opportunity, through intense discussion, to evolve an idea so that it can be its most complete version. Brainstorming is a form of collaboration. Most inventors are great collaborators. They relish in the creative process of sharing ideas and concepts with like-minded people. Architects, artists, designers and scientists -- people with the gift for invention and design -- revel in the process of brainstorms.

It is not uncommon for an invention to evolve into something very different from its initial design. It can either be something more complex or something more simple than initially intended. This is alright because the entire point of the invention process is to create a product. Whether the original idea was kept or not, the significant point is, the idea was transformed into an actual product.

Part of the process of invention is to distinguish between an idea that is going to be functional and an idea that is not going to be functional. If in the process of inventing a product it has been deemed not practical, then it is alright to stop. There are many reasons for discontinuing production of an invention. Some of them are: there is a similar product in the market already, or it is not a marketable product, or it is a product that has yet to be useful today. If this happens then by all means stop developing it.

Most of the time, a product is invented because its creators want to monetize it by selling it to the market. Unless the inventor is a very rich philanthropist, with the resources to finance whimsical projects, then that is the only way a very specialized product will see the light of day. Most of the time, if an idea for a new product shows it does not have marketable value then it does not get invented.

A good example of a novelty product is a solar powered submergible amphibian car. There is no market for it today. Very few people will be interested in a car that can be driven underwater. The only reason this kind of vehicle will be invented is if a wealthy pro-environment person who has a fascination for easy amphibian transport commissions a car designer to create it for him. Although not impossible, building this sort of car is rare because of its very impractical nature.

Another example is the parachute. When Leonardo Da Vinci thought of the parachute way back in the 16th century, air travel was not yet invented. So during his time the parachute would not have been a practical invention because it served no purpose during that time. Part of being a successful inventor is having the ability to distinguish between ideas that are relevant to the times. You do not want to waste your time inventing something you cannot sell.

During Da Vinci's time invention was purely a creative exercise. Almost like a hobby. These days the art of invention is almost always business driven. Since the industrial revolution inventing things became much easier. Today most businesses are the result of the rise of the art of invention during that time. It is the reason why some of the most successful businesses in the world now are driven by invention and innovation. It can even be said the wealth of the people in the 21st century is on account of the rise of the art of invention.

If you have the talent for creative thinking and are able to transform your ideas into tangible products and gadgets, by all means, keep on doing what you are doing. Stay on that invention course. Who knows you may have the makings of the next Steve Jobs or Henry Ford.

INVENTION: THE CONCEPTUAL WAY

As I said earlier, inventors are some of the most creative people in the world. They have an ability to see beyond the limits of reality and imagine possibilities outside of the realm of the usual. Inventing objects require a very vibrant imagination. It also requires some or a lot of technical knowledge to translate imagination into reality. The best and most successful inventors are the people comfortable in the conceptual world and relish the opportunity and time to think and imagine.

All invention stem from imagination. Without the capacity to dream and imagine it is not possible to invent new and extraordinary things. Invention starts with a creative mind and the capacity to imagine. This is the reason that inventors are the best dreamers.

a. **Play**: As an inventor always encourage the inner child in you to experiment. Be curious. Try out new things. Make play a part of who you are. Play exercises the imagination and opens opportunities for learning. Play is a quality that is essential in encouraging inventiveness. Another important aspect of play is the ability to daydream. Daydreaming is the ultimate form of imaginative play. It frees the mind from thinking of the mundane and practical things in life. It is the ability to step out of the reality into a world of the imagination. Daydreaming stretches the mind towards the realm of fantasy. If you look at the history of the best inventions throughout the century, some of the best of them were first imagined while their inventors were daydreaming.

b. **Re-envision**: Another quality of inventors is the ability to see things in a new way. Leonardo Da Vinci was known for examining the very first design of the clock and added a few features that made it even better. To be inventive, it is necessary to look at an existing product or design and imagine a better version of it.

Do not be discouraged if new ideas do not come immediately. Sometimes it takes time. Inventors have many 'Eureka moment' stories where solutions to problems pop up in their minds many days or months even years after encountering the problem. A good example is Einstein. Although he was working on the theory of relativity for years, without any meaningful breakthroughs, he finally had his Eureka moment in a dream. He woke up one day with the answers to the theory of relativity. So give it time. Re-envisioning a better version of a product does not always come at the moment you are looking at the product. Let it percolate in your brain and in time you will find the answer.

c. **Insight**: The ability to see a situation, person or thing with a deep, intuitive understanding is insight. Inventors have a great capacity for seeing through things and discerning their true meaning, purpose or potential. Inventive insight is essential in invention because it sees potential and possibility in the unusual or even the accidental. A good example is the metallic plastic sheet used in

Flat screen TVs today. The material is actually a design accident.

What happened was too many catalysts were added on to regular plastic sheets during a routine production process. It resulted in the plastic sheets acquiring a metallic reflective color. An insightful inventor looked at the metallic sheets and realized it could be used for improving and developing new products such as display screens, lights and wallpaper.

Insight is the ability to see possibility. A trait that is common in many successful inventors and idea-makers.

d. **Exploration**: All invention is exploratory. It is the creation of something new. New means it is unexplored territory. The outcome can be predicted but remains unknown until the process of invention is completed. Because invention is exploratory the result can either be a success or a failure. Regardless of the outcome, all forms of invention go through a process of development. The goal of course is for the invention to be a success. When it is not. No need to fret. The investment of time and resources poured into the process of invention is not all that lost. The knowledge gained from the experience of experimentation and exploration will be useful for future use.

e. **Improvement**: Inventors can be counted on to make improvements on almost anything. They can examine a product or process and suggest ways to make it a better product or process. They can improve the value of a product by adding new features. They can make any product work more efficiently. Inventors have the knowledge and skills to streamline the invention process so that producing a new product is more cost efficient and less time consuming. Product improvement is a challenge that great inventors like to take on.

Give yourself time for a little self-examination. Ask if you have these qualities of inventors. Do you like play? Do you have the ability to re-envision things? Are you excited with the possibility of making improvements on products that you encounter every day? Do you enjoy exploring possibilities? If the answer to all the questions is yes, then you have the makings of an inventor.

Nurture these qualities in yourself. In the process of developing that new product you have been daydreaming about for the past six months, make sure to try to re-envision it to expand its possibilities. Also remember to apply insight by looking at your product with a deeper understanding of its nature. Try to see further into its potential. And always, always have a mind for improvement. Make it second nature to ask, 'how can this product be further improved?'

Remember an idea is merely a possibility, a kernel of thought in your mind, until you invest the time, resources, and energy to put it through the process of invention. Challenge yourself to take an idea to the next level. Invent it.

Chapter 2: Innovate

Innovation is the byword of the 21st century. Almost every successful company today has embraced the value of innovation as a means of achieving continuous growth. To innovate is to make improvements on a product, service or process. Or sometimes, it is to introduce an entirely new product or service that outperforms older versions. Innovation is a continuous process. It aims to keep on making improvements on existing products and services to ensure relevance and competitive advantage.

There are many ways to innovate. Some of the most effective ways are:

1. **Find inspiration from other inventions:** Henry Ford visited a meat packing plant one day and had an insight about the production line

process. He realized that he could use the same in his production of Ford vehicles. When he used the production line process to make his vehicles he was able to reduce assembly time and production cost so much that he was able to build more vehicles than he initially expected.

Copying an idea and applying it into your product, process or service is a form of innovation. The key is to be very observant about what ideas are useful to your business or not. Make a habit of observing the new and improved ways people are doing business. They may contain ideas that will prove profitable for your business.

2. **Ask customers:** There is a reason many business establishments get customer feedback on a regular basis. Customers are the best source of information for innovation. Since they use the product or experience

the service, your business offers. They can provide valuable information on ways to make improvements. Listen to what they have to say and choose the ideas that are truly innovative.

3. **Listen to complaints:** When a client complains about a product or service then they are pointing out an area where innovation is required. Always make it a point to address a complaint and ensure that the new and improved product or service is a better version of the older model. Listen to complaints they usually lead to innovation.

4. **Observe customers:** The denim jeans company Levi Strauss observed their customers ripping their jeans as a fashion statement. In response, they started offering pre-ripped jeans in their stores. An astute business decision that allowed the company to create a niche market that resulted in millions of dollars in earnings. The

same is true for the ketchup company Heinz. They observed how people were using their product. The result was an unusual realization. They tilt the ketchup bottles upside down to make it easier for them to pour the ketchup. In response, they decided to produce upside down Heinz ketchup bottles. As soon as the bottles hit the supermarket shelves, the shoppers picked them up.

To innovate observe the people using the product. Usually the way they use the product or respond to it can give insight on how to make the product or service more innovative.

5. **Ask staff:** Next to customers and clients, the business staff is the best source of information for innovation. They are in the frontlines after all and will always have the best ideas for improvement. Encourage staff members to embrace a culture of innovation in the workplace. Conduct

regular staff meetings where they can share ideas for innovation.

6. **Eliminate:** When the people at Dell developed their business they did so with an eye for eliminating the computer store. They built their business in a way where customers could go to them directly and make a purchase. The same with Sony and their Walkman. They innovated by eliminating the record and speaker function normally found in radios. They replaced it with the headphone. This innovation made the Walkman a purely music and audio listening device.

Amazon used the process of elimination to innovate their business model. They had a specific goal, to make bookstores obsolete. When they started selling books through their website and guaranteed delivery of every purchase anywhere in the world. They were able to achieve their goal. To innovate, think of ways of streamlining the process so that unnecessary steps are eliminated.

7. **Combine:** Try looking at your product and ask what could be added to it so that it can be improved. When wheels were added into suitcases, innovative thinking was at work. When the camera was added to mobile phones, the inventors and product developers where being innovative. Look around and observe, you will find some form of innovation in almost every product you encounter. To innovate think of adding a new feature.

8. **Plan:** Make innovation a goal. Push your business to list innovation as a target that needs to be achieved. Add it into the timeline of your business. Inform all business staff of the goal to innovate. Tell them the specific timeframe for the expected innovation. Move your business to

meet the target. If you motivate everyone in your business to innovate then they will.

9. **Brainstorm:** Focus Group Discussions are so popular in many companies for a reason. It brings together a diverse set of people from within the company and sometimes from outside the company and forces them to discuss ways to make innovations on a product. The diversity of information gathered in FGDs almost always leads to innovative wisdom the business can adopt.

10. **Check patents:** Ask the nearest government patent office for list of new patents, expired patents and all existing patents relevant to your product. The idea from the expired patent can be applied to the product you are developing. Since it is already expired, the information in the patent can be used by anyone.

The ideas from the new patents can be checked against your product and service to make sure that they are not similar. If they are similar then you may not be able to move forward with developing your product. Ideas from existing patents can be a source of inspiration for innovation. Checking patents is a great way to make sure you are developing an original product.

11. **Maximize or Minimize**: Aim to be different not better. Sometimes customers gravitate towards something that is cutting against the grain instead of competing against many who are similar. Minimizing a product or service is a form of innovation. Maximizing a product or service is also innovation. Don't aim to be the best. Aim to be different.

12. **Collaborate:** Georg Jensen is a jewelry making company that greatly

admired the work of star architect Zaha Hadid. They saw that her aesthetic sense could be translated into modern jewelry design. They got in touch with Zaha Hadid and asked if she would be willing to collaborate and design her own line of modern jewelry. She agreed. Now the Zaha Hadid jewelry line at Georg Jensen is one of the company's best sellers. Collaboration leads to fresh ideas that spark innovation.

13. **Sponsor a contest:** In the same vein as asking clients and your business staff for feedback, sponsoring a contest that allow innovators and inventors to submit improved version of a product is a great way to discover innovative ideas. Offering a prize is sure to motivate people to participate. Just make sure the goals and expectations are clearly set at the start of the competition. If you do that, more often than not, the public will

surprise with a product that will be very profitable for your business.

14. **Explore possibilities:** Sit down with your team and ask the 'what if' questions. Pretend for example that there are no limits and no boundaries to inventing a new product. See where that thinking will lead to. Sometimes the best ideas are arrived at when there are no constraints attached to it. Innovation is about exploration and exploration is about crossing boundaries and finding out what is on the other side of the norm.

15. **Observe the competition:** If you are a big company it is wise to take note of the small fish in the pond. These are the businesses that are in direct competition with you. Often the smaller businesses are hungrier and are more innovative than big companies. Observe them. See what new innovations they are cooking up. Better yet, if your business is big

enough, buy the small fish in the pond. Staff your company with the most creative and innovative people you can find.

16. **Open innovation:** Unilever and P&G, two giant multi-national FMCG companies have a policy of taking in and purchasing novel products from developers. Other than internal product development, they have opened their companies to new product invention and product innovations from independent product developers. They understand that although their companies are the leaders in the field, with massive resources poured into product development, they do not have exclusive rights to innovative ideas.

17. **Outsource:** The same with open innovation instead of waiting for independent product and service developers to bring you their innovative ideas. See if you can

outsource your product development. There are tons of design companies who offer the service. And will be more than happy to collaborate with client companies. Also for fresher ideas, try asking the local university to let their product design students to have a shot at product innovation. There is also the option of hiring a start-up. Some companies are brave enough to even crowd source their new product. The point is, do not limit product innovation to the four walls of your business.

18. **Triz:** There is a system called Triz you may want to consider. Engineers and product designers love this product development system. Triz is used to solve systems and design problems. It is a kind of toolbox. It contains a variety of methods that can be adopted to untangle contradictions. It answers tricky questions like, 'how to build a product that operates at lightning speed using cost efficient renewable energy?' Give it a shot. It may

surprise with new ways of seeing your business.

19. **Consult history:** It has been said that the popular 21st century dating game called 'speed dating' was actually inspired by an 18th century Victorian dance tradition. During the Victorian age at costume balls, men make an appointment to dance with a lady and she lists the appointment down. In a similar vein, Speed dating allows men to date women at a given timeframe within the duration of the event. Looking back at history and identifying old practices that can be tailored to modern 21st century methods are a great way to innovate.

20. **Use a product for a new purpose:** De Beers is the world's leading diamond company whose core business is industrial diamonds production. These are the diamonds used for industrial purposes. In the process of developing industrial

diamonds, De Beers stumbled upon designing a diamond engagement ring. Seeing the potential of the product, they sold it in the market. Now De Beers is one of the leading designers of engagement and wedding rings in the world. Because they followed through on an innovative idea they were rewarded by a niche market that is proving very profitable for the business. One that allows them to earn billions of dollars every year.

21. **Social networks:** Some of the most creative, forward thinking people are found in social media. These are the people who have their own blogs and talk about the virtues of certain products and services in their site. Having an eye towards these mavens, whose opinions thousands of followers value, can be a good idea. Engaging them to get their feedback on your product and service can be an effective way to arrive at innovative breakthroughs.

When searching for the next innovative idea always remember, when it comes to innovation no one has exclusive rights to the best ideas. The key is to always have your eyes and ears on the ground and keep consulting with people. The next best innovative idea is out there waiting to be discovered. Keep looking.

Chapter 3: Create a prototype

The two most important steps in the process of invention and also innovation are designing and building a prototype. A prototype is a three dimensional version of a design. It is the physical representation of a product. It is tangible and functional according to the purpose it was designed for. Depending on the design, prototypes can be as basic as piecing together everyday objects in the house. Or it could be a bit more complex and sophisticated with actual mechanisms that operate and power it.

Whether it is a basic design or a complex one, all prototypes are initial ideas of the product. They are used for the purpose of refining product design. Prototypes are early models intended for testing. Creating a prototype is very important in the process of invention.

Here are some of the reasons prototype creation can be very valuable:

1. **Tests for functionality:** On paper an idea can look great. It can seem convincingly useful and effective. However, when a prototype is created often the design flaws are detected. Design and functionality issues start to crop up. A prototype is a great way to test if the product is working according to the intended design. It also provides the opportunity to refine the design so that functionality issues are addressed.

2. **Tests performance of materials:** As I said earlier, prototypes check for design flaws. For example, a prototype built using a combination of metal and glass. It could happen, while testing for functionality, the glass parts make the product brittle and fragile. The inventor then can decide to use sturdier materials such

as carbon fiber, industrial grade plastic or crystal.

Without a prototype, design glitches can only be identified on the finished-ready-for-market product. By then it would be too late to make design changes. Sometimes it is a costly decision not to create a prototype. If the design is changed after product completion then more money needs to be invested in making an improved version. Creating a prototype is always a wise business decision.

3. **Opens opportunity for refinement:** The process of testing and checking the design and functionality of a prototype opens the opportunity for refinement. That is the very purpose of a prototype, to be able to evolve a design into a better version of itself.

4. **Defines the product:** At the end of the process of refining a prototype, when the product has gone through extensive tests and checks, the product can be finally defined for what it is. Since the product features have already been identified then they can be added into the product description along with functionality. Defining the product is very important in the product marketing stage.

5. **Creates credibility:** Investors, marketing companies, and licensing companies take a person carrying a prototype into a meeting far more seriously than designers or inventors who do not. A prototype tells investors that the inventor is serious about their business venture. That the inventor has invested valuable time, energy and resources in creating a functional prototype to allow them to see how viable their idea is. Important people and companies are likely going to give the

entrepreneur with a prototype the support required to allow a new product to compete in the market. A prototype gives the inventor and entrepreneur credibility. It is a symbol of competence.

DOWNSIDES OF PROTOYPING

Most people will argue that developing a prototype is not as beneficial as some would like to think. They consider it an unnecessary step that requires investment in capital and resources. Naysayers argue that it is an extra step in the production process that can be skipped. Here are some disadvantages that should be considered:

1. **Too much development time:** Since prototypes are meant to be tested and checked for design flaws and functionality, product development gets stuck in the prototype stage. End-users, designers, and developers get so engrossed making refinements and

improvements on the prototype they end up wasting time. Instead of moving forward to actual production, the product is stuck. All because people can never be satisfied with making improvements.

2. **Causes confusion:** When the final product is produced, it is always based on the approved prototype. However, there are instances where certain features are removed in the process of production. Since end users expect to receive the same version of the approved prototype they approved. It can be confusing and disappointing to receive a final product that does not have the same features as the prototype. This is the reason it is absolutely necessary to be engaged not just in the process of building the prototype but also in the process of producing the final product. The design team or inventor needs to have final say on any adjustments on the product design during the production stage.

3. **Too much focus on parts:** Prototype inventors and designers have a tendency to nitpick on the individual parts of a prototype. They like to focus on refining a certain feature or element of the prototype. They devote all their time and energy in finding ways to make that part look better. This is a fault in designers. Although refining parts is important it is more important to develop a prototype looking at it as a whole. Time is wasted refining parts when it should be refined as a complete system.

4. **Expensive:** Some inventors and designers prefer to skip the prototype making process to save on cost. They think it is an unnecessary expense that eats up on valuable financial and human resources. The argument is, 'why waste time building a prototype when the same time and resources can already be used to create a complete and fully functional

product?' A product ready to be sold to the market. Some inventors argue that prototypes can be expensive.

In the process of creating your product take time to sit back and gauge if going through a prototype making process is a necessary step. Some products are simple enough not to warrant making a prototype. It can go directly to production. If you are confident about this then it is always an option to skip building a prototype. However, if you think that the product design merits a testing and refining stage. By all means, build a prototype. Go with your better judgment on this and make the investment to build a prototype first.

DEVELOP A PROTOTYPE

Convinced about the value of prototypes? Then go ahead and start building one. There are very many ways to approach the process of creating a prototype. It will depend first on the simplicity or complexity of the design.

Creating a prototype of a Valslide for example is less complicated than creating a prototype of a biofeedback device. The former only requires a piece of plastic with a handle on the one side and a textured surface on the other. While a prototype of a biofeedback device is a far more complex design since it requires electronic functionality.

Depending on the design of the product, here are some of the possible approaches to creating a prototype:

Build it yourself: As mentioned earlier, if the design of your product is simple enough you can build it yourself. Look around your house and gather everyday items that will allow you to piece together your design. Remember there are no rules to creating a prototype. It is the process of experimentation and exploration. The goal is to allow you to see what materials work best to build the product. So go ahead. Give it an initial shot based on initial design sketches. If you are already in the prototype development stage then it is likely you have already thought the product through. And it is ready for production. The next step is to build it.

Collaborate with a professional: For more complex product design it is a good option to collaborate with or hire a professional prototype developer. Yes, there are people and companies who are in the business of creating prototypes. They will have the creative manpower and technical resources to allow you to transform your product design into an initial three dimensional version.

Do a bit of research in your city or community about professional prototype developers. You are likely going to find several companies or freelance individuals who provide the service. The great thing about hiring a professional is they have tons of experience to contribute - always good when creating a product. It is best to approach these professionals with design sketches prepared. It should contain detailed information on functionality and design that will guide them in developing the prototype.

The more specific you are about your prototype design the better for the pros to build according to expectations. Since you are hiring a professional, make sure your idea truly deserves the financial investment. It only makes sense to hire a professional if your design is too complex for you to build it yourself. If it is a simple design, best to work on it yourself.

3D Printing: New technology is another option for building a prototype. An example of this is 3D printing. Much like when we print a document using a printer, a 3D printer allows a designer to input a design on to a computer connected to a 3D printer. Then print said design as a 3D object on the 3D printer. 3D printing is now a very popular technology, so popular that most prototype developers already have their own 3D printers. There are many versions of 3D printers available in the market. They range from very expensive industrial grade 3D printers to $30 Do-It-Yourself models that anyone can purchase.

If you are a prolific inventor you may want to invest in your own 3D printer. Make it easy to print your product designs from the convenience of your home or office. However, for more complex designs it is best to let the professionals handle 3D printing your new product. Also, reminder 3D printing will only print the shell of your product or the parts of it. If the prototype requires electronics and mechanization, they will have to be added to the 3D printed prototype for it to function.

Whichever prototype development process you choose always remember, this is the exploration and experimentation stage. The entire point of building a prototype is to be able to check if the product is viable. If it is then do everything you can to build a better and more refined version of the initial design. Do it right soon you will be on your way to moving forward with mass production of your invention.

Chapter 4: Patent

The next step in the process of selling a new invention is to apply for a patent. This is a very important step in the invention process especially if you already have a fully functional prototype. A patent is a form of intellectual property right that protects your idea or new technology from being copied and sold by other inventors. It is a set of exclusive rights given to an inventor of a new technology. For the duration of the patent, usually twenty years, the owner of the patent has exclusive rights to the idea, the design and the technology. No other person or company can use or produce the same product or design other than the patent owner.

If other people use it other than the patent owner they are obligated under the law to ask permission from the patent owner. The patent owner then has a right to ask for financial remuneration for the use of the patent. The patent owner can also set the terms and limits for use of the patent. These are just some of the reasons why it is very important to get a patent for every new idea, design or prototype that you develop as an inventor or entrepreneur. It is necessary to protect your original ideas – your intellectual property – and there are laws in place to help you do exactly that.

Every country has a patent and intellectual property rights department. This is the government agency that grants patents to anyone who requests for them. Inquire with your local government and ask where you can get a patent for inventions and new technology. The government department assigned to issuing patents will have a list of rules and regulations for patent application. They will also be able to provide guidelines on terms and limits for every kind of patent. Make sure to get all relevant information about patent application in your area. It is information that will allow you to get approval for your patent.

Although important, a patent is not a mandatory step in the invention process. There is an option not to patent a new design or new invention. And it should not deter you from producing it for market consumption. For more information on the advantages and disadvantages of getting a patent, here are some of the reasons to consider.

ADVANTAGES

1. When you have a patent for your invention you have the right to prevent people from copying your invention. The public is also constrained by law from manufacturing your invention and selling it without your permission. When they do you are allowed to file legal charges against them and demand financial settlement.

2. As a patent holder only you are authorized by law to have commercial rights to your invention. Only you can market and sell your invention to the public.

3. As a patent holder you decide the purpose for which it will and can be used. You decide how it will be used and who will use it. You have exclusive rights to determine the fate of your invention.

4. Once you have a patent for your invention you have the option to sell the license to anyone who is interested. If a business, for example, wants to produce your invention to sell to the market then they can approach you for a license to do so. This means you can charge for royalty fees. The terms of which you can agree with

the person or company purchasing the license to the patent.

5. As the owner of a patent you can set the price for the purchase of the patent. If a business wants to purchase the patent, meaning transfer ownership of the patent from you to them, you as owner of the patent and the technology that it represents get to name the price.

6. If your business has competitors that develop the same products, a patent is protection. Since you have sole ownership of the idea, the product design and technology indicated in the patent, none of your competitors, according to law, can manufacture, develop, or sell the same product that you have a patent for. Only your business can produce and sell it.

7. Since patents allow people to have access to information about new technology, other inventors in many industries benefit from the knowledge shared and transferred. Some of the industries or sectors that have benefitted from patents are medical science, biotechnology, drug chemistry, computers, etc.

8. Patents are a source of inspiration and encouragement for gifted inventors. It is a way of rewarding them for investing their talent, energy, resources and time in the creation of technological advances. The financial reward is a motivation for many more inventors to keep on building and creating new products and technologies.

DISADVANTAGES

1. Because patents are made available to the public for the

specific purpose of sharing knowledge, your competition or anyone in the world can take your idea and use it for their purpose. All they have to do is to wait for the patent to expire and they are free to steal the ideas behind the invention, manufacture it, and sell it to the public without paying for a license of use or asking permission from the patent owner.

2. Since the details of the patent are made available to the public, your competition can study it and make a better version of the product. Since the knowledge is shared with the public, it is possible for other people to make improvements on the technology and create a better version.

3. Applying for a patent is an investment in itself. It is a very long process. It usually takes several years before a patent is

approved. It requires an investment in time and financial resources. It takes many rounds of presentations before a patent is issued.

4. There are also patent processing fees involved. This means it requires a certain amount of financial investment as well. Sometimes the cost of the patent fees amount to more than the production of the product. Also, if you would like to ensure that your invention does not get produced in other countries. It is necessary to go through a patent application in every country you want your patent protected. That requires a sizeable investment in financial resources.

5. Sometimes another person will file an infringement complaint and that will mean you should be prepared to explain your patent

and defend it. You must be able to prove to the patent office that the idea is in fact an original and that you were the first one to have thought of it.

Getting a patent is discretionary. Being the inventor of a new technology, you decide whether there is value in applying for a patent. How important is it to you to protect your idea from being stolen by another person or business? A large enterprise will almost always apply for patent to protect their innovative products. They will have the personnel to manage it and the finances to support the patent application. If you are an individual entrepreneur, it may not be as easy to go through the process of patent application. Almost not, wise to invest the financial resources to have exclusive rights to your idea.

Think it through before going through the process. Some products may be too simple to require a patent while others can be so innovative and have breakthrough quality that it is necessary to cover it with intellectual property rights via a patent. In the US, the government office in charge of authorizing patents is the United States Patent and Trademark Office. Inquiring with them about additional information on patent application will enlighten you on details about the process.

Chapter 5: Sell and Market

The final step in every inventor and entrepreneur's journey is selling their original invention or innovative product in the market. Often it takes years to allow an idea to germinate and fully transform into a viable product that will resonate with the mass market. As mentioned in the early chapters, for a new gadget or product to be competitive it needs to go through so many layers of testing before it gets the seal of approval for mass production. But when it does and you finally have the crates of inventory to prove it then it's time to move on with selling and marketing it to its intended market.

CREATE A SALES PLAN

Since you are the originator of the product, it is likely that you have your market identified already. The market is comprised of the people you think will need the product enough to purchase it. It is necessary to identify the market because the marketing and sales plan will be tailor fitted to their needs. The more specific the target market the more effective the sales and marketing strategy will be.

For example, it will be too broad to identify the market as women in the US. Considering it is too broad a spectrum of many types of women. It will be difficult to sell any product to them without categorizing the type of women. A more specific definition would be – Asian women, 35-45 years old, with college degree, married with children, health buffs. It is easier to create a marketing plan with a specific target market in mind.

The next step is to develop a sales plan. This is a crucial step in the process of selling your product. It will be the roadmap to achieve sales targets. A typical sales plan should include the following:

Sales Goals: In order to achieve sales goals it is necessary to be specific and also reasonable about the sales numbers that needs to be achieved in a set period of time. For example, 100 units in the first 3 months to individual buyers or 500 units in the first 3 months to retailers in the region are sales goals that can be achieved in the timeframe provided. In stark contrast with sell 500,000 units within the year. That is too broad strokes and does not detail to whom the units will be sold to and where. Setting manageable and specific sales goals gives you a roadmap for selling your product within a set time period.

Sales Activities: The sales plan should also include specifics on sales activities that you plan to engage in. Sales activities are the strategies you will employ to sell the product to the market.

They are tactics for making a sale. For example, joining a bazaar or craft show is a sales activity. Putting up a website to showcase your products and sell it directly to consumers anywhere in the world is a sales activity. Printing a brochure containing all the products you are selling with detailed information about functionality and features and giving them to retailers is another form of sales activity. Usually the product and the target market will determine the best form of sales activity to employ. So think it through and choose the ones that will ensure you reach your target market and make a sale.

Target Accounts/Market: Be very specific about your target market so that you can tailor fit the sales plan to them. If they are retailers, for example, it may be necessary to schedule product presentations to their acquisition team so that they can make bulk orders for their retail stores. If the target account or market are individual end-users. The best way to reach them may be via product website, E-Bay or Amazon. Make sure you know who to sell your product to.

Timeframe: A schedule is great motivation for achieving your goals. Make sure to set a timeframe on every sales activity and goal. In business, time is equivalent to profit. The faster you are able to dispose of units or items, the more profit you are able to turn around after investment. Set a timeframe so that you can also monitor how effective your sales and marketing activities are. If you are lagging in your set schedule then you can find ways to address the reason for the delay. Gives you reason to think of solutions that will increase the pace of sales.

In the process of selling to your market, observe the sales activities that are proving effective. Make sure that you continue to adopt these activities in your sales strategy. If possible you can improve on it in order to get more sales.

Whatever you do make sure to keep the sales strategies that work. Every entrepreneur dreams of being taken on by a large retailer and have them sell their product in their stores. However, most retailers will look at the sales track record of a product before they decide to add it to their shelves. Having a great sales record will allow you to sell to large retail companies that order products in bulk to sell to a nationwide or even global market.

DEVELOP YOUR MARKET

For an individual inventor and entrepreneur new at selling a product it is best to start small in developing the market. There are a lot of things to learn in the process of building the product market from the ground up. In the business world consider yourself a hatchling and in order to fly you have to take baby steps at expanding your wings and getting comfy at flying. This means sell your product direct to end-users first.

Before it became a multi-million dollar enterprise, the creator of Tupperware products hosted Tupperware parties in their home to the women in their community, the women who are the end users of the product. These Tupperware parties allowed them to get valuable feedback from their market. Information they used to improve on the products they were selling. Tupperware sold great products that filled the need of many homemakers in the US. Soon it was being sold in the entire country via Tupperware parties. It is always a good idea to sell direct to end users in the process of building your market.

These days an effective way to reach end-users is through the internet. Creating a product website that allows the target market to see the product, read up on its features and functions and make a purchase, is a good way to develop the market for your products. And in much the same way as the Tupperware parties, it is also possible to get valuable feedback from end-users through the website. The upside of making your product available on a website is you are able to reach a bigger market than say merely hosting a party in your community or neighborhood. If you do not want to make the investment to build your own website, then you can always sell your product via E-Bay or Amazon or the many retail websites available in the web today.

The advantage of selling to end-users first is you are able to test how popular your product is to the target market. Number of sales is proof that the product has a market. And that it is worth selling to the public. End-users are also a source of vital information that allows you to make further refinements on the product based on user experience. Make sure you listen to end-user feedback and make the necessary adjustments. They are your clients after all and speak for your target market. Selling to end-users first prepare you and your product for the big leagues – the retail stores.

Again even with selling to retail stores it is reasonable to start small. Scheduling a presentation with local, independently owned store owners should be the next step after direct selling to individual customers. Since you will be talking to the store owner directly, the decision maker that will decide whether to purchase your product or not, getting a sale is more likely to happen. Local stores often appreciate new products or gadgets. They like the idea of selling novelty products that are not yet available in the large retail chains. These products make them more competitive. Again selling to a local store owner provides valuable information that will be useful when you eventually sell to the big retail chains.

EXPAND TO BIGGER MARKETS

The entire point of starting from the ground up with selling and marketing your product is for it to establish a strong and solid sales record. This is possible only if it has a large end-user market base and is being carried by several small local retail stores. This is proof that the product sells. Information that you can bring to the big retail stores to convince them to carry your product in their shelves. This means you can now aim to sell your gadget or product to nationwide mass retail stores such as Wal-Mart and Target. This expands the market to millions of end-users who shop at these giant retail stores.

When you do get to this stage remember the sale is just the beginning. Because of the massive inventory often required. Mass retail stores after all need a big inventory so that they can make the product available in all their stores nationwide. Your business should be ready to handle the inventory demand. Time to make another plan, make sure your business is able to manage product handling, rollbacks, returns, advertising and all the other additional systems required to meet the scale of demand by giant retail companies.

Here are a few pointers on how to make a successful pitch to large retail companies:

Identify correct buyer: Large retail stores have specific buyers for every category of FMCG products. In order to sell the product to them the first order of business is to identify the buyers who will authorize purchase of the product. In order to get names, try inquiring with the headquarters of the retail store directly. If that does not work then talk to distributors of the retail store and ask them for information on buyers to get in touch with. If they are not generous with the information then it may be necessary to invest in hiring a manufacturer's representative to make the proper introductions with the buyer.

Preparation is key: Once you are able to schedule a meeting with a buyer then make sure you have a decent presentation prepared. It is always good to look professional. The best way to do that is to arrive prepared for the meeting. Complete with product sell sheets and also sample products the buyer can check.

Identify the target: Impress the buyer with data about their product line and how the product you are offering will be a good addition to their inventory. This is especially effective if your product is a novelty product that the store is yet to carry. Dazzle them with the products sales record. If possible compare it with the sales record of a competitive product. Understanding the nature of the retail store and aligning your product to their standards is a great way to get a sale.

Watch out for special programs: On occasion mass retailers such as

Target enforce a policy where they authorize their local store managers to sell local products in their stores. This means it is not necessary to visit the main office of the mass retail store and make a pitch to the buyers of the store. If you want to sell to the local store only because your product is local specific. Watch out for special concession programs that allow it. Whatever you do keeping any eye on the activities of mass retail stores is always a good habit for expanding your business.

Patience is a virtue: The main goal of most entrepreneurs with a product or gadget to sell is to have them available on the shelves of mass retail stores located around the country. After all, it is the best way to reach the product's end-users. They give access to millions of product consumers increasing the possibility to earn more profits. And if the product proves to be very useful and popular so much so that it becomes a staple weekly purchase of the average shopper. It translates to more orders and continuous production. This leads to business success. So it pays to be patient. It may take years before the mass retail stores agree to sell your product but when they do the wait will be worth it. Be patient.

Conclusion

Thank you again for purchasing this book!

I hope this book was able to help you in forming a solid plan to developing the next thing that would make the world a better place.

The next step is to get out there and just create.

Finally, if you enjoyed this book, then I would like to ask you for a favor, would you be kind enough to leave a review for this book on Amazon? It'd be greatly appreciated!

Thank you and good luck!

www.ingramcontent.com/pod-product-compliance
Lightning Source LLC
Chambersburg PA
CBHW060414190526
45169CB00002B/889